Deere & Company's Early Tractor Development

by Theo Brown

1953

AF692303

Contents

 • International Standard Book Number 1-887446-04-4 •

Theo Brown must be considered one of John Deere's most productive farm equipment engineers. Born in Worcester, Massachusetts, on April 19, 1879, he was christened Theophilus, a name he rarely used and was known by few of his associates; but it was the name he signed on his patent application papers.

Theo was granted his first patent just two years after graduating as a mechanical engineer from Worcester Polytechnic Institute in 1901. The patent was for a manure spreader end-gate. He was employed at the time by the Richardson Manufacturing Company of Worcester, which built Kemp & Burpee Manure Spreaders. In 1911, Deere & Company bought the Kemp & Burpee spreader line and moved production to the Marseilles Company in East Moline, Illinois. (Marseilles was a subsidiary of Deere & Company, and eventually became the John Deere Spreader Works.) Deere hired Brown at that time. While still employed by Richardson, he designed and patented a manure spreader with "The beater on the axle." This spreader, which John Deere introduced, really put the company in the manure spreader business. In 1912, Theo was named superintendent of the Marseilles factory.

In 1916, Deere transferred Theo to the Plow Works to manage that factory's experimental department. In this new position, he was involved in the engineering of plows and several other products — including development of an early motor cultivator. During World War I, he helped design a wooden "combat" wagon which was sold to the U.S. Army for use by the American Expeditionary Forces in France.

Because Deere was a basically "decentralized" company, each factory had its own product development department. Factory engineers generally tended to be somewhat protective of their "turf," and often secretive about new product plans. Theo Brown recognized the dangers of that approach, and convinced top management that if Deere was to succeed in the tractor business, there must be more coordination and cooperation between factories, and especially between tillage implement factories and the tractor factory at Waterloo. When he was elected to Deere's board of directors in 1923, Theo became the company's "chief engineer" and had a legitimate interest in product development across all product lines.

In 1925, Deere president C.D. Wiman committed the company to the development of an "All-Crop" Tractor that could compete with the newly introduced McCormick-Deering Farmall, a product of the International Harvester Company. The end result was the introduction of the John Deere Model "C" in 1927, the Model "GP" in 1928, and the "GP" Wide-Tread in 1929. Theo Brown was the chief advocate of the 3-row implements designed for the "C" and "GP". Unfortunately, farmers preferred 2-row and 4-row equipment. The "GP" Wide-Tread answered that demand, and actually introduced several excellent features at the time, but design compromises initially made with the "C" carried over into the other "GP" models, and they were less than overwhelmingly successful.

The Great Depression would make it difficult to immediately proceed with sweeping improvements in tractor design, and these were difficult years for Theo. He wrote in his diary, "An experimental man needs enthusiasm to do good work. This depression affects original thinking." But, as soon as possible, development of new tractors and implements was underway. Theo Brown was very much involved in the engineering effort that brought forth the Model "A" Tractor in 1934, and the Model "B" in 1935. That same year, the American Society of Agricultural Engineers (ASAE) awarded him the Cyrus Hall McCormick Gold Medal for "outstanding achievement in the field of engineering applied to agriculture."

In 1937, C.D. Wiman put Theo in charge of developing a new line of "baby" combines. Two years later, the John Deere Nos. 10, 11, and 12 Combines were introduced; and replaced a year later by the improved Nos. 10A, 11A, and 12A Combines.

For more than another full decade, Theo continued to stay closely involved with the development of continually improved John Deere agricultural equipment. He retired from his engineering responsibilities in 1952, but remained as a director of Deere & Company until 1954 — bringing to an end 43 years of outstanding service to Deere, and more than a half century to the farm equipment industry.

Theo's professional colleagues regarded him as a man "of eminent fairness, unfailing friendliness and ever-smiling good naturedness." Brown's sharing attitude is, perhaps, best reflected in his own words from the mid-thirties...

"I have endeavored in the experimental work with Deere & Company to develop people. Instead of taking a large number of patents myself, I have been interested in getting others to think independently by giving them problems to work out themselves. In this way they have had the satisfaction of taking out many patents in their own names.

He remained active and interested in Deere & Company throughout his adult life. It was in 1953, at age 74, that he wrote the book *Deere & Company's Early Tractor Development*. At age 81, he appeared at Deere Day in Dallas for the introduction of the New Generation Tractors. It was a grand affair attended by some 6000 implement dealers and branch managers from the United States and Canada. *The Dallas Times Herald*, in an article appearing August 31, 1960, noted that Brown held 158 patents and that "He is credited with more inventions for tractors and other agricultural implements than any other person who has ever been employed by the Company."

During his last 35 years, Theo maintained a summer home on Mountain Road, Princeton, Massachusetts. He died at a hospital in nearby Holden on June 30, 1971, at age 92. His wife, Elise, preceded him in death by three weeks.

FOREWORD

The material used in this report was gathered from many sources, and pieced together to form a connected and comprehensive account of the earliest activities of Deere & Company in tractor design and development. An endeavor has been made to record the thinking that prompted these activities, to record what was done, and also to tell why these various early activities were discontinued.

This report will reveal that many features in tractor design, which today are considered both desirable and novel, were incorporated in these early tractors of Deere & Company — as, for example, Sklovsky's one-piece cast iron body; Melvin's integral power-lift plows; Dain's transmission, changing from low to high speed without stopping the tractor; reversing the tractor for certain operations (Melvin's tractor), and the single-row motor cultivator.

In the early days of tractor development, creative thinking went beyond the ability to transform such thinking into practical use. Had today's engineering knowledge and manufacturing technique been available at that time, many of the novel ideas would have proved commercially successful.

On March 14, 1918, Deere & Company purchased the Waterloo Gasoline Engine Co. This report covers the activities of the Company up to that time, and is divided into five sections which are devoted to as many distinct projects by Deere & Company engineers.

The Appendix contains information about men mentioned in this report. Acknowledgment is given to the following for their assistance: Nathan Lesser, Walter Silver, Joe Dain, Leonard Neighbor, Fred Kirby, and Miss Martha McCormick (Elmer McCormick's sister). Much information was gathered from the records of Deere & Company Board of Directors' meetings, from the files of the Deere & Company Patent Department, and from my personal diaries.

Theo Brown
1953

Original text edited and format modified to correspond with the layout and photographs used in this publication.

Chapter 1 — The Melvin Tractor

There was a time when the opinion of agricultural leaders differed as to the practicality of gas tractors — when many farm implement manufacturers wondered whether they should build gas tractors or attempt to ignore them.

For that reason, this record of the early participation of Deere & Company in the development of gas tractors logically begins with March 5, 1912, for on that day the Deere & Company Executive Committee passed a resolution which was to influence the Company's entire future. It read:

> "RESOLVED: That in view of the inevitable future use by farmers for diverse purposes of gasoline and kerosene tractors, and especially since the trend is to use them in connection with implements, particularly plows, it seems vital to the interests of this Company that serious cognizance should be taken of the situation, and that through its experimental department, the personnel and talent of which shall be increased, if necessary, a movement to produce a tractor plow should be started at once having in view constantly, that the success of the same would be enhanced if not assured, were it possible to divorce the tractor from the plow and to thus make it available for general purposes."

On May 23, 1912, Mr. Mixter reported that Mr. Max Sklovsky had fully investigated the various tractors, with conclusions as follows as to the best tractor in each class:

Heavy Tractor Class — Aultman-Taylor
Smaller Tractor Class — Avery
Motor Plow Class —Hackney

On July 1, 1912, C.H. Melvin was transferred from the Experimental Department of the John Deere Plow Works to the General Company and given the assignment of designing and building an experimental tractor plow.

A room in the Plow Works was provided for Mr. Melvin, and there he built the tractor shown in the photographs. Only one Melvin tractor was built.

Records show that a total of $6000 was expended for this purpose. There are no records to be found that refer to the performance of this outfit in the field, but Joe Dain remembers that the results were disappointing both as to field performance and keeping the tractor from breaking down.

Max Sklovsky also had ideas and produced drawings for a motor plow, dated July 15, 1912, and August 24, 1912, but such a tractor was not built.

The general layout of Mr. Sklovsky's drawings has a semblance, in some aspects, to Melvin's tractor. Both Melvin and Sklovsky must have been influenced by the design of the Hackney motor plow.

In 1914, it was decided to stop any further development work on the Melvin motor plow. The transition from the heavy, slow moving tractors pulling 12 to 16 plows to a small tractor that could pull three plows was well underway.

Some of the first efforts to design a small tractor resulted in mounting the plows under the tractor frame midway between the front and rear tractor wheels. The Hackney motor plow as an early example of such a construction and Melvin followed the idea in his tractor. This type of tractor was not successful. ■

Above: *The Melvin Tractor of 1912 was the first ever to bear the John Deere name. The two holes at the front were for hooking up the direction guidance device.*

Page 5: *Melvin's tractor had two driving seats, and centrally located controls, so it could be driven in either direction. For drawbar work, the tractor ran in the opposite direction than for plowing. All tractor wheels ran on unplowed ground, but the guidance device had one furrow wheel to assist the operator in maintaining the desired cut.*

Chapter 2 — The Dain Tractor

In 1914 Deere & Company was skeptical about getting into the tractor business, but realized that further development work should be carried on. This hesitancy of Deere & Company can best be expressed by quoting from the Company records. These records also give the story of the Dain Tractor.

"May 27, 1914. Mr. Dain was asked to report to the Executive Committee whether or not a tractor could be built to sell at about $700 and in the meantime to suspend work of development until his report is made.

"June 9, 1914. Mr. Dain said he was working on a plan which he would submit as soon as completed. It was Mr. Webber's idea that we should decide very soon whether or not a light tractor could be built to sell to the farmer for about $700. Mr. Dain further stated that the tractor he is planning would operate a 3-bottom plow.

"June 24, 1914. 'RESOLVED, that the preliminary work of designing an efficient small plow tractor be continued under the auspices of Mr. Dain and the Experimental Department.'

"September 8, 1914. Mr. Dain reported that $6,000 had been expended in connection with the work done by Mr. Melvin and further stated he thought he could build a light tractor for experimental purposes for about $3,000. Approved.

"Dain Asked to Continue

"February 13, 1915. Mr. Dain has developed a small experimental tractor and it is questionable whether we should enter into the manufacture of this machine. 'Now, Therefore, Be It Resolved, That Mr. Dain continue his work with the tractor now built until he considers it perfected, and that it be ascertained at that time the amount we have invested, for the purpose of fixing a basis upon which we would sell the same to a separate company for its development and sale. Also, that Mr. Dain advise the basis upon which his transmission can be used in said tractor.'

"March 9, 1915. It was the sentiment of the Board that Mr. Dain should continue his work in connection with the tractor, give it a thorough trial in the field, and continue its development; that the present was not the time for Deere & Company to decide whether it should go into the tractor business or not; that it might seem wise later and it might not; that the wise thing for the Company to do was to watch the development of this business, and also watch and develop the tractor, in order that the Company might be ready for any emergency.

"Mr. Dain reported that up to date the amount expended on the present tractor was about $2,890.

"May 1, 1915. The general situation in reference to the new tractors was discussed.

"Webber Requests Field Trials

"Mr. Lourie stated that the tractor people are selling plows right on their tractors, and that it is going to interfere with the wheel plow trade. Mr. Webber stated that he had been pretty well satisfied to leave the matter the way it was because the situation in all territories was not alike. What he meant was, that we did not know what we had in that tractor yet, until it had done a lot of work. He stated that Mr. Dain was going to revise the next one he builds, and is going to make some improvements. He thought that the tractor ought to be kept at work to find out the weak places, and that the next one should be sent to the Minneapolis territory where it can plow 300 or 400 acres. Mr. Lourie thought that we should get out at least five tractors and put them in various conditions.

"Mr. Butterworth stated that we had repeatedly told our bankers that we were not in the tractor business and were not going into it. Mr. Lourie stated that the idea had occurred to him that it might be feasible to have an independent company organized as a selling company, and to let our factory here produce the tractors and the sales company sell them.

"Mr. Webber stated that, with the danger ahead of losing plow trade, as suggested by Mr. Lourie, we ought to be prepared to keep moving. He stated that he was very much impressed with the opinions of Mr. Schutz and Mr. Hornburg on the Dain tractor. They stated that they had never seen a tool or an implement that impressed them so favorably as that machine. They believed it had great qualities and would be a tremendous success if put on the market.

"Other Tractors Heavy and Clumsy

"Mr. Webber thought that we could go out and say that Deere & Company is going to put this engine on the market; that it is an engine that will pull three plows, and two in

Page 7: *This photograph, probably made in February 1915, shows the first experimental tractor built for Deere & Company by Joseph Dain Sr. after development of the Melvin Motor Plow was discontinued. The all-wheel drive was an important advantage (over other contemporary tractors) of this and subsequent Dain tractors.*

difficult plowing; and that it is a good engine, or Deere & Company would not make it. He thought that the bankers had very good reasons to be scared about gas tractors, but that they had in mind the Rumley, the Emerson and the Hart Parr people. They have made a heavy, clumsy machine and have made a failure, and have not made any money. They think it is something we are going to sell on two or three years' time, while we are talking about something that is entirely different and that will go to the farmer. He thought that the question to be decided in the end was whether this machine or any little engine is going to be an economical thing for the farmer to buy. That we do not know as yet.

"The following resolution was unanimously adopted:

> "'RESOLVED: That without attempting to at this time define our final policy toward the tractor business further than the resolution passed at the Directors' meeting held February 13, 1915, but with the view of trying out more thoroughly the Dain Tractor, Mr. Mixter be requested to cooperate with Mr. Dain in order to hasten the try-out of the present machine and to determine upon the improvements that should be made in it, with the authority to build from three to six of the revised experimental tractors to be thoroughly tried out this Summer and Fall.'

"Mr. Dain Makes Full Report

"December 16, 1915. Mr. Dain presented the following report on experimental tractors.

"'We have, to date, made three machines — two with friction transmission, and one with a gear transmission that in all other respects follows the same general plan as the friction machines.

"'The first tractor, weighing about 3800 pounds, was finished last winter and tested in the shop for strength and power. We obtained in these tests, by means of artificial traction, a steady drawbar pull of 5000 pounds, on slowest transmission speed. As soon as the weather permitted we put the machine in the field and obtained a drawbar pull of over 3000 pounds with our natural traction.

"'In plowing we first tried the machine in five-year-old clay sod, pulling the three 14" bottoms at two and one-half miles per hour. After plowing all spring, we pulled two binders in heavy oats at an approximate speed of three miles per hour. This fall we sent the tractor to Mr. S.H. Velie's farm near Kansas City, where it plowed about 85 acres. The conditions here were about the most difficult we have had as the land is hilly and the soil in the draws a very heavy clay that was harder to plow this fall than usual owing to the heavy summer rains followed by hot, dry weather.

"Minor Mechanical Troubles

"'No accurate record has been kept of the amount of work done by this first machine, but it will easily amount to over three hundred acres. We have been very well pleased with its performance, considering that, as it is entirely different from any other tractor on the market, we did not have anyone's previous experience to guide us. The first chains were too light, causing some breakage, but this has been entirely overcome by putting on

Above: *Mr. Dain's second tractor. His third experimental tractor, completed in December 1915, was practically identical to the one shown here except for the transmission; a positive gear-driven transmission was used instead of a friction transmission.*

Note the center location of the gearshift lever, the offset seat and steering wheel, brake pedal, and drawbar.

Page 9: *Mr. Dain's second experimental tractor is shown here plowing wet gumbo soil in the Red River Valley near Winnebago, Minnesota. The cost for 80 acres of 6-inch plowing (three 14-inch bottoms) was 59 cents an acre, including the operator's time at 30 cents an hour.*

a little heavier chain. We broke two front ratchets, which was easily remedied by making them of a little better material. The front wheel drive has proven very satisfactory, and has demonstrated to us that this is a most desirable feature. The gears in the rear wheels show considerable wear. We had at first considerable trouble with the motor overheating, but since putting on a radiator recommended by Long Mfg. Co. and a larger size fan we have had no further trouble of this character.

"Machine No. 2

"'This tractor, weighing about 4000 pounds, was finished this fall, and after plowing about fifteen acres near Moline was sent, at the request of Mr. Webber, to Winnebago, Minn., where it plowed eighty acres at a cost of fifty-nine cents per acre, counting the man's time at thirty cents an hour. The soil here was of heavy black gumbo and was in poor plowing condition owing to the almost continual rain — several days we plowed, the neighboring farmers thought it too wet to plow with their horses. We pulled three 14" bottoms 6" deep at two-and one-half miles per hour. The farmer plowing in the next field was using five good horses on an 18" sulky — he tried using six horses on a New Deere gang, but the draft was too heavy for them.

"Machine No. 3

"'This is the geared transmission machine, and has just been finished. It is identical with the other machines, excepting the transmission, and instead of a worm and gear and internal gear drive to the rear wheels has a chain drive — this change necessitated by the transmission we are using, and from the results obtained by the other two machines we believe this type of drive to be satisfactory in every respect.

"'Judging from the tests we have made in the shop with this tractor, we believe we are in a position to recommend it for serious consideration and comparison with the friction transmission machine.'

"March 14, 1916 Mr Mixter read to the Board the following telegram from Mr. Dain, in reference to the tractor being tried out in Texas:

"'San Antonio, Tex., March 13, 1916

"'Deere & Company, Moline, Ill.

"'Have followed tractor closely for two weeks. Conditions extremely hard and rough. Absolutely no weakness in construction. Gears, chains, universals, in fact all parts in good condition. Tractor has travelled near five hundred miles under extreme load. Change speed gear a wonder. I recommend to the board that we build ten machines a once.

"'Joseph Dain.'

"Mr. Mixter stated that what little difficulty they have had with the tractor was due to poor material; that he thought Mr. Dain felt that there were some details yet that could be worked out to get the tractor on a manufacturing basis; that this tractor in Texas is the first one of that type that has been tried out — the first gear transmission. The friction transmission has been essentially dropped. He also stated that we could probably manufacture tractors in existing factory capacity of an expenditure for new machinery, pattern, tools, jigs, etc. from $25,000 to $50,000.

"Board Approves Further Development

"The following resolutions were passed:

"'WHEREAS, Mr. Joseph Dain has developed several small tractors which have been experimentally tried out, the latest machine prov-

Above Left: *This quartering front view of the second tractor* (pages 8 and 9), *built in 1915, shows a rather crude early version of what would later become known as "The tractor ahead of its time."*

Page 11: *The Marseilles Company factory in late 1917, during the time the "Dain" tractor was being tooled-up for production. Elmer McCormick supervised the manufacture of 100 tractors; he left Marseilles to become Chief Engineer of the John Deere Waterloo Works.*

JOHN DEERE
MOLINE ILL
MARSEILLES CO
MARSEILLE

ing successful in Texas for a sufficient time to demonstrate its practicability from a general standpoint, and

"WHEREAS, It is desirable that this machine should now be reconstructed to meet such minor questions as have developed in Texas, and to make it suitable for economical manufacturing and then turned over to a manufacturing organization.

"RESOLVED, that Mr. Dain be requested to reconstruct said tractor with such modifications as may be deemed wise by him and his associates, preparatory to economical manufacturing;

"FURTHER, That the Marseilles Plant (now the John Deere Spreader Works) be directed to take up this tractor work with the object of getting it on a manufacturing basis as a possible part of their regular line;

"FURTHER, That about ten machines be built by the Marseilles Company, at the earliest practical date;

"FURTHER, That both Mr. Dain and the Marseilles Company be directed to vigorously continue tractor development work.'

"McVicker Designs Engine

"June 13, 1916. Mr. Dain reported that the Marseilles Company had started to manufacture five machines, and that one was being built here at Moline, which would complete six of ten tractors authorized at the last meeting of the Board. One of the tractors had been shipped to Minot, North Dakota, and was doing very satisfactory work. A new motor is being developed as rapidly as possible, Mr. Dain stated, with the assistance of Mr. McVicker of Minneapolis. The difficulty experienced with the motors on the market is the inaccessibility of all the parts. This feature is being overcome in the motor which Mr. McVicker is working on at the present time.

"On July 13, 1916, Mr. Mixter reported that 'McVicker design of engine should be complete about July 17 and in view of preliminary work already done one engine should be complete about August 17.

"'We can count on four to six machines of the Dain type with Waukesha engines for shipment in August, providing design of rear axle and transmission is decided on within ten days. We can count on a modest tryout of one or two engines of the McVicker design this fall.

"'Original cost figures on this machine were $736 (1.25 steel) — $761 (2.25 steel). These figures are based on paying $200 for the motor. The writer has been over the detailed labor figures on the latest estimates of cost and believes that the machine as today designed with the motor figured at $200 can be built for $600 (1.25 steel) cost of manufacture.

"'This means, judged in the light of other goods of our manufacture, that the farmer should pay $1,200 for the machine. This is somewhat higher than has been considered admissible for a three-plow tractor. It is the writer's belief, however, that an all-wheel drive will ultimately be the tractor the farmer will pay for.'

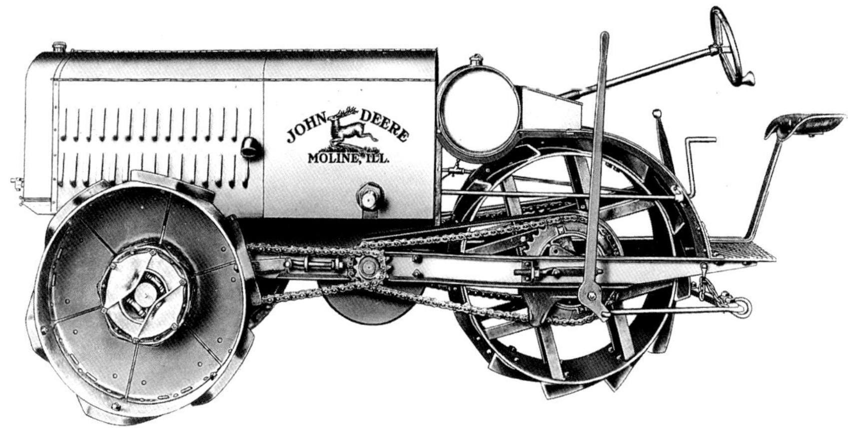

"Extensive Field Tests

"September 12, 1916. Mr. Webber stated that three of the Dain tractors were in the Minneapolis territory. One was at Minot, and had been there since spring. It had done quite a lot of breaking, pulling two 14" plows in hard ground, and had performed very well. One was at Huron, South Dakota, and had been used for running a threshing machine, but was now plowing. One had also been sent to Fargo, but had just arrived.

"Two of the best Minneapolis men have been detailed to visit these machines several times between now and freezing-up time, and Mr. Webber and Mr. Velie expect to go with them, if possible, and to keep close track of their work.

"The 1916 Minneapolis Report

"Mr. Dain read a report from the Minneapolis house on the operation of the tractors in their territory, as follows:

Above and Page 13: *The photograph on page 13, taken of a not yet fully assembled tractor, was used by a Deere & Company artist to develop the image above used in the tractor's advertising brochure. The twin rows of hood louvers were replaced by a single row of longer louvers (on each side) when regular production commenced.*

"'November 13, 1916

"'Deere & Webber, Minneapolis, Minn.
"'Gentlemen:

"'Mr. Molstad and the writer after having visited Dain tractors at Aberdeen, Fargo, and Minot each three times, have the following to submit:

"'We are firmly of the opinion, providing the clutch collar and bevel pinion sleeve trouble can be eliminated, and not taking into consideration the power plant or the question of belt power, that we have in the Dain very much the best tractor on the market, principally for the following reasons:

"'Best on the Market

"'1st. Because of its all-wheel drive, minimum weight, and maximum traction.

"'The all-wheel drive makes its light weight possible and gives the maximum traction, the advantages of which are that the tractor will

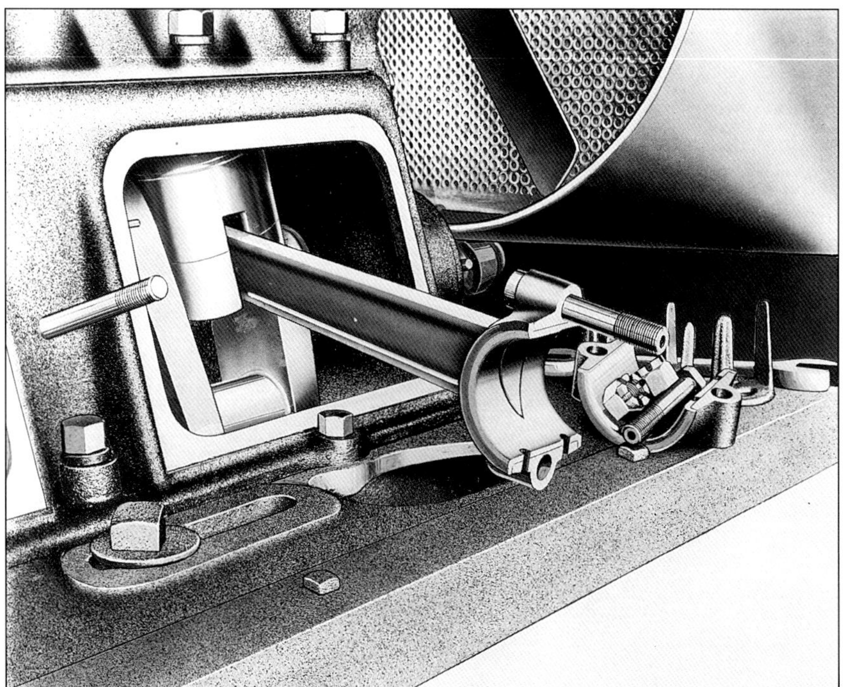

go thru more difficult conditions without miring down than any other tractor, and in the event it does mire down, by uncoupling the plows, it will always pull itself out, which other tractors cannot do.

"'Because of its light weight it packs the ground less than other makes — which packing is generally conceded to be a serious objection.

"'2nd. Four chains being used to drive the wheels divides the strain on the sprockets and chains, making the chain drive, in our judgment, two or three times more durable than gears, and much easier for the farmer to replace and adjust.

"'Another big advantage of this feature is that there is no noise. The noise made by the gears and the exhaust, on many tractors, especially after they have been used for some little time, is almost unbearable.

"'3rd. Because of being able to change speed without clashing the gears.

"'4th. Being able to change speed without stopping saves time, trouble and gasoline, and prevents a loss of momentum.

"'5th. Because of its foolproof simplicity and accessibility.

"'The Question of Power

"'Question of Power: We are firmly of the opinion that no one will *ever* build a tractor that will prove satisfactory for plowing without having a power surplus of 33-1/3% or 50% over and above the amount required to do the work. In other words, in building a tractor to pull 3 bottoms in stubble or 2 in breaking, the engine should have power enough to pull from 4 to 4-1/2 bottoms under average conditions. This would mean that the tractor would have to be marketed so as to insure the user not pulling more than 3 bottoms in stubble or 2 in breaking — by making him sign an agreement to that effect, the agreement stating that in case of its being violated it would invalidate the Warranty.

"'As further precaution would recommend that a cast plate be put

***Left:** Two removable plates on the side of the four-cylinder, in-line McVicker engine, developed with Dain's input, allowed access to bearings, rods, and pistons.*

***Page 15:** Four John Deere All-Wheel-Drive Tractors are lined up for a 4th of July parade in 1919, each pulling another product of the Marseilles Works.*

HUBER EXPRESS CO.
HUBER EXPRESS CO.
Marseilles Works.
JOHN DEERE

on the tractor somewhere stating that the tractor is guaranteed to pull only 3 bottoms in stubble or 2 in breaking.

"'A tractor, like a man or horse, cannot work to its limit continuously and last.

"'The greater the number of cylinders that are used in the motor the less the amount of surplus power required, because the more cylinders there are, the more the motor is like a steam engine.

"'Gray... Waterloo... Heider Tractors

"'At Aberdeen we saw a $2,150 6-plow Gray tractor, and its owner. It was pulling one of our No. 2, 4-bottom Pony Engine Plows. We asked the man why he was not pulling 6 bottoms. He stated that he wore out one tractor in a short time by pulling all he could with it, and he made up his mind that the only way a tractor would stand up was for it to have enough power so it would 'play' with the plows, so to speak.

"'We also saw a Waterloo and a Heider tractor at work. Neither of them had plowed 500 acres.

"'The Heider had plowed about 400 acres and it was in very bad shape. The noise it made was terrific. Everything about it seemed to be loose. It was about ready for the scrap heap.

"'Service Problem

"'Mr. Harrington, of the Fargo Implement Company, has sold 17 Waterloo tractors — 10 this year and 7 last. He stated that if he had 17 more out it would keep him and all of this men busy doing nothing other than look after the tractors; and he has about concluded not to sell any more of them on account of the general effect it would have on his business — regardless of the precaution he takes when selling them.

"'He [Mr. Harrington] stated that if a customer has trouble and they all seem to have an undue amount, he will blame the Fargo Implement Company just the same, and when the farmer is ready to buy a gang plow the probabilities are he will go to their competitor and get it.

"'On the other hand, Harrington said, if he had not sold these tractors he probably would have sold only 4 or 5 of our Pony Plows instead of the 17 that he sold this year.

"'Good Tractor Needed

"'This goes to show that our dealers need a tractor badly, but they must have a good one.

"'We must not only have plenty of power, as indicated above, but

Above Left, Above, and Page 17: *The John Deere All-Wheel Drive was actually put on the market and many ended up in the hands of farmers. This one is pulling a spring-tooth harrow and a rock-weighted, wood-constructed drag to prepare the seedbed.*

must have an extra-strong, heavy, well-built motor, with very liberal bearings.

"'We do not believe that the Waterloo tractor, except under very favorable conditions, can pull more than one 16-inch sulky plow and stand up.

"'As to the number of bottoms the Dain tractor should pull we have recommended, and the factory has been working along these lines we believe, pulling three 14-inch bottoms under favorable conditions, and two in breaking; we are wrong in this! Our tractor should be strong enough and have power enough to pull three bottoms in stubble, under almost any condition, as the majority of farmers when purchasing a 3-plow tractor expect to be able to do better and deeper plowing than they could with horses. In fact, a good many tractors are sold at times when plowing conditions are unusually hard for horses.

"'Three-Bottom Power

"'Purchasers are not going to be satisfied, and the sales will not be repeaters, unless the Dain will pull three bottoms in stubble generally.

"'Further, under most all sod-breaking conditions in our territory it is just as hard or harder to pull two 14-inch bottoms as it is to pull three 14-inch stubble bottoms under hard plowing conditions.

"'Comparing the work of the three tractors referred to herein, the Aberdeen tractor is the only one that proved satisfactory to the user — although they had some carburetor, magneto and fan belt, and other minor troubles. The motor in this tractor, so far as we know, had no connecting rod trouble. The user of this tractor and a neighbor of his were so well pleased with it that they will buy Dain tractors as soon as they are put on the market. Although this tractor has the smallest motor of the three, viz. 4-1/4 x 5-3/4, it was working under very favorable plowing conditions.

"'The work done by the Minot and Fargo tractors, particularly the Fargo tractor, was, we would say, nearly twice as severe — so severe that considerable motor trouble was experienced. Although there was no connecting rod trouble with the Fargo motor, they plowed only about 200 acres.

"'An extra-heavy, strong motor throughout is just as essential as is the surplus horsepower it seems to us.

"'Salability of the Dain Tractor. There is no question in our minds that the Dain tractor will easily outsell all others on account of its having so many plainly apparent features of merit, such as the all-wheel drive, etc., enumerated in the forepart of this letter.

"'Quality Should Come First

"'We think the matter of price should be forgotten for the present. Go ahead and build the tractor — first class all the way through, using extra-good magnetos, carburetors, etc., as well as making it extra-good in other details, and when that is done if the price must of necessity be $1,500, to market them profitably, let's sell them for that. We, of course, could not sell as many at $1,500 as at $1,200, but the factory probably could not make them, to begin with, any faster than they would sell at a good, fair, round price — and above all we are favorable to marketing them at a price that will net the dealer a good profit, so he will push their sale and can afford to look after them.

"'Further, with reference to the price: We know that the Dain as Deere & Company will probably build it, will figure stronger per

Above and Page 19: *A 30-inch diameter steel pulley, with 8-inch face and spoke hub, could be attached to the left side of the tractor to produce a belt speed of 2190 feet per minute.*

JOHN

pound than does the Bull or Waterloo. We cannot see why the Dain tractor should run into a cost that will make it necessary to sell it for a price as high as $1,500. It weights only about 4000 pounds. The Bull and Waterloo each weigh about 5000 pounds, the former selling for $685, the latter for $750. The Gray tractor (4-plow) which is an expensive, well-built machine, in fact a good deal better than the Dain, weighing 6000 pounds, 50% more than the Dain, sells for $1,650 cash.

"'Tractors and Plows

"'In considering the matter of price we must remember the more tractors we sell the more tractor plows we will sell.

"'Yours very truly,

"'Geo. P. Schutz (Signed)'

"September 12, 1917. Mr. Webber stated that the tractor situation was changing in this respect — the demand is for a tractor that will pull three plows — and he believed it will be the one that will pull two, three, or five. He thought we could sell the Dain tractor for $1,500. He said that the Dain tractor was all right; that it was a 3-plow outfit, and that we would probably have to eventually make a bigger one.

"Mr. Webber reviewed in detail the performance of the Dain tractors in the Minneapolis territory, which were very satisfactory.

"RESOLVED, That Mr. Dain be authorized to continue his negotiations with the thought in mind of buying not over 100 of the Dain tractors outside with the view of continuing the development of the tractor line, the purchase to be made on a fixed price basis.'"

Mr. Dain's Untimely Death

On October 31, 1917, Mr. Joseph Dain died in Minneapolis following a trip to Huron, South Dakota, to investigate the performance of the Dain tractors. George Schutz accom-

Left: *Joseph Dain Sr. was founder of the Dain Manufacturing Company, which in 1911 was sold to Deere & Company. Dain became a vice-president of Deere & Company, and a member of the Board of Directors. He was in charge of designing the John Deere All-Wheel-Drive Tractor. He died in 1917, before the tractor went into brief production.*

Page 21: *The All-Wheel-Drive Tractor could handle a three-bottom plow, but cost about double the amount of other tractors of the period.*

panied Mr. Dain on this trip to South Dakota, and under date of November 7, 1917, reported as follows:

"November 19, 1917. The subject of Dain tractors was considered and the report of November 7, 1917, signed by Mr. G.P. Shutz and approved by Messrs. D.D. Velie, D.D. Webber, and F.R. Todd in reference to these tractors, was read and spread upon the minutes.

"'November 7, 1917

"'Subject: Huron, South Dakota, Dain Tractor Report

"'The undersigned, with Mr. Joseph Dain, Sr., spent two days, October 24 and 25, at Huron, South Dakota, investigating the performance of three Dain tractors.

"'Mr. Velie's signature appears on this report, he concurring in what is stated herein, having visited at Huron to see these tractors in operation a few days prior to the above dates.

"'F.R. Brumwell Tractor

"'The first tractor we visited was on the farm of our agent, F.R. Brumwell. This tractor had plowed 110 acres; had harvested 260 acres, and had drawn five wagon loads of stone at a trip on two different occasions to Huron, about 12 miles distant.

"'We saw this tractor pull one of our No. 5 NA214 3-furrow Pony tractor plows, plowing six to seven inches deep, several times around the field. To test the tractor, we set the plows several times to run eight- to ten-inches deep, and it handles the load fine, with no engine knock or undue slippage of traction wheels.

"'Plowing conditions: Somewhat harder than normal. Soil: Heavy, black loam, subsoil being quite hard below six inches. This tractor had been operated by two different hired men of average ability.

"'Mr. Brumwell is very enthusiastic over the Dain tractor. He believes it to be much better than any other he has seen.

"'He had a few suggestions to make for refinement of some of the details, which among other suggestions, will be found further on in this report.

"'In 1916, we loaned Mr. Brumwell a Dain tractor, equipped with a Waukesha motor, which tractor was returned to the factory last winter, overhauled and equipped with a new motor. The 1916 machine gave such good satisfaction that he has bought three tractors equipped with the new motor, one for his own use, and two which he has sold, being the three tractors referred to herein.

"'J.J. Hoffer Tractor

"'This man had plowed 70 acres 8-1/2- to 9-1/2-inches deep with his tractor, pulling one of our No. 5 NA214 3-furrow Pony tractor plows. This was two- to three-inches deeper than the ground had ever been plowed before.

"'Plowing conditions: Difficult, heavy black loam soil, quite dry.

"'We followed the outfit around the field several times, at one end

Above: *Stability of the three-wheeled tractor was undoubtedly a consideration, especially so in light of the fact that plow counterforces were applied to the narrow end.*

Page 23: *At least one John Deere All-Wheel-Drive Tractor was responsible for developing farm land.*

Specifications

Rating .. Draw-bar, 12-horse-power; belt, 24-horse-power.
Speeds .. Two forward and two reverse. All speeds direct. No intermediate gear used. High, 2-5/8 miles per hour; low, 2 miles per hour. Reverse speeds same as forward.
Weight .. 4,600 pounds.
Length .. Wheel base, 6 feet 2-3/4 inches; over all, 12 feet 6 inches.
Height .. 4 feet 9 inches.
Width .. 6 feet 4 inches.
Wheels .. All traction. Rear, 40 inches in diameter, 20-inch face; front, 36 inches in diameter, 8-inch face.
Steering Gear .. Worm and sector.
Transmission .. Exclusive design. No gears to shift in changing speeds.
Change can be made from low to high, or vice versa, while under full load without stopping the machine.
All gears are cut from special heat-treated and tempered steel, mounted on ball and roller bearings and enclosed in an oil-tight case.
Chains are of highest grade steel and are of the roller type.
Sprockets .. Cut from heat-treated and tempered steel.
Motor .. Size, 4-1/2 inches by 6 inches. John Deere design. All bearings extra large.
Large hand-hole plates through which the bearings may be adjusted or the pistons removed.
Cylinder .. Cast in block and ground to size. Detachable head. Extra large water jackets.
Pistons .. Extra long and ground to size.
Piston Pin .. Extra large, 1-1/2 inches in diameter, fastened in the piston. Bearing in connecting rod head.
Connecting Rods .. Drop-forged from open-hearth steel and double heat-treated.
Crank Shaft .. Three bearing. 2-1/2 inches in diameter.
Drop-forged from open-hearth steel and double heat-treated.
Bearings .. Connecting rod, 2-1/2 inches in diameter by 2-7/8 inches long; front main, 2-1/2 inches in diameter by 3-7/8 inches long; center main, 2-1/2 inches in diameter by 3 inches long; rear main, 2-1/2 inches in diameter by 4-7/8 inches long.
All bearings bronze back with special babbitt metal lining.
Lubrication .. Force feed from mechanical oiler, supplemented by a splash system.
Governor .. John Deere design, built-in.
Cooling .. Water circulated by positively-driven centrifugal pump of extra large capacity.
Fan .. 19 inches in diameter. Positive gear drive.
Carburetor .. 1-1/4-inch Model M-2 Stromberg.
Air Cleaner .. Bennett.
Ignition .. K. W. high tension magneto with starter coupling.
Pulley .. Steel, split hub, 30 inches in diameter, 8-inch face. Belt speed, 2,190 feet per minute.

Series B101-D.&Co.-1919

The
John Deere
All-Wheel-Drive
Tractor

going through quite a strip of sod on high speed and without changing the depth, the engine handling the load splendidly.

"'This farmer was very enthusiastic. Said it was the best tractor in his community; that his neighbors liked it; that one of them owning a Case, purchased last year, said the Dain was a much better tractor than his.

"'Frank Lyon Tractor

"'This tractor purchased only recently, and therefore he had plowed only a few acres with it. We were not able to see it in operation, a bolt having gotten into the transmission gears and broken two cogs out of the high speed gear.

"'Mr. Lyon, however, was highly pleased with this tractor, the feeling that the breakage was nothing for which the machine was to blame.

"'1917 Minot Tractor

"'In addition to the three Dain tractors, with the new motor, sold at Huron, Deere & Webber Company's agents at Minot sold one, which has plowed some 200 acres with one of our 14-inch 3-furrow Pony tractor plows, 100 acres of which plowing was done with this plow as a 2-furrow with breaker bottoms.

"'Mr. Schutz visited this tractor after the farmer was through using it, and examined it carefully. He found it in first-class condition. While he did not see the owner, the Minot Implement Company stated that he was very much pleased with the tractor, having experienced no trouble with it whatever.

"'Other Tractors Visited

"'We saw two small 4-cylinder single-drive-wheel Case tractors, the users of which both told of having been stuck at times due to lack of traction power. These Case tractors are very complicated and inaccessible. We also noticed that the large main wheel driving gears were badly worn.

"'We saw two made by the Avery Company. One as a 2-plow tractor, with a 2-cylinder opposed engine equipped with two carburetors, one for kerosene, one for gasoline. This tractor is extremely inaccessible, and so narrow it would tip over very easily.

"'The other was a 2-row motor cultivator, which the owner said was worthless, owing mainly to the engine being no good and having insufficient power. He used it but a short time, finishing up this cultivating with horses.

Left: *Back and front of the original 4-page sales brochure, which measured approximately 8-3/4-inches high by 15-inches wide when unfolded.*

Page 25: *Where are they now? Just as with nearly 30,000 Waterloo Boy Tractors, most of the John Deere All-Wheel-Drive Tractors disappeared long ago. The rear hood of this one had already been set aside when the photo was taken in 1920.*

"'We also saw a Bates Steel Mule tractor and interviewed the owner. It was not in operation at the time. We had experienced a lot of trouble with it. This tractor does not look practical to us.

"'For want of time we were not able to visit many other makes, but we were impressed by the few we did see with the fact that many tractors lacking in important essentials are being marketed.

"'First Engines Too Light

"'The three Dain tractors equipped with Waukesha motors, loaned to customers in the Minneapolis territory last year, to be tried out, one each at Fargo, Minot, and Huron, which plowed 300 to 400 acres each, proved entirely satisfactory in all essentials, except that the engines were too light for tractor work and lacked power.

"'Some trouble also was experienced with the thrust collar for fly-wheel clutch, and bevel pinion drive shaft bearings giving out.

"'We have had none of these troubles whatever with the four 1917 Dain tractors put out this year, viz; the one at Minot and the three at Huron, the engines having stood up and given perfect satisfaction; no difficulty of any kind having been experienced with the clutch thrust and bevel pinion shaft bearings, which were redesigned and equipped with ball and roller bearings.

"'The 1917 engine is a splendid one, having much more power, but no more than needed, extra strength so it will stand up, and is made very accessible. So, by using heavier chains, which have already been tried, and heavier driving sprockets, to measure up with the stronger motor, we feel sure it will take the Dain tractor out of the experimental stage.

"'CONCLUSIONS

"'We believe we have in the Dain tractor one that will create a sensation and prove to be a big seller, for the following reasons:

"'1. The three-wheel drive. Will go through more difficult conditions without miring down than any other tractor, and in the event it does mire down, by uncoupling the plows it will always pull itself out, which other tractors cannot do.

"'2. Maximum traction with minimum weight. Because of its light weight, which is evenly distributed on all traction wheels, it packs the ground less than other makes, which packing is generally conceded to be a serious objection.

Above Left: *One of the few survivors, the tractor that was salvaged by F.L. Williams of Cordova, Illinois, was missing the hood and correct radiator when this photo was taken many years ago. Note the belt pulley.*

Above and Page 27: *The serial number plate and part of a bullet-ridden carcass are about all that remain of this tractor in Nebraska.*

"'3. Transmission running in oil, insuring perfect lubrication at all times.

"'4. Change of speed made without clashing of gears.

"'5. Being able to change the speed without stopping the tractor saves time, trouble and gasoline, and prevents loss of momentum.

"'No 'Unbearable Noise'

"'6. Four chains being used to drive the tractor divides the strain on the sprockets and chains, making the chain drive in our judgement two or three times more durable than gears and much easier for the farmer to replace and adjust. The four chain drive precludes the possibility of a tie-up through breakage of any one of the chains or sprockets. The chain is noiseless. The noise made by the gears and the exhaust on tractors generally, especially after they have been used for some little time, is almost unbearable.

"'7. Motor: More durable, being heavier, for its bore and stroke, than any other tractor motor. Valves so made and timed as to limit supply of gasoline, making it foolproof, so operator cannot overload engine with gasoline and tear motor to pieces.

"'Simple and Accessible

"'8. Because of its foolproof simplicity and accessibility throughout. No other tractor on the market begins to approach the Dain in

Above: *Dain's tractor design reached the market place.*

this respect. No other tractor on the market has the motor so made that the pistons and connecting rods can be removed without tearing motor down.

"'The pump and magneto are separate from the camshaft and can be removed without taking the motor down.

"'The cylinder head can be removed to take out carbon and grind valves without tearing engine down.

"'The transmission is easily gotten at, with plenty of room to work, by simply removing transmission cover.

"'Finally, we firmly believe the machine, generally, is perfected to a point where it can be successfully manufactured commercially, and we strongly recommend that you proceed with the construction of at least 100 as soon as possible.

"'Yours very truly,

"'Geo. P. Shutz

"'Approved:
"'C.D. Velie
"'C.C. Webber
"'F.R. Todd'"

Board Orders 100 Dain Tractors

"The following action was taken (by the Deere & Company Executive Committee November 19, 1917):

"RESOLVED, That Mr. Clausen be requested to proceed with the manufacture of 100 tractors of the Dain type, using for that purpose such members of the organization as are available and such outside assistance as it is advisable to obtain, provided, however, that the personnel of the organization called into this branch of the business shall be decided upon after conference with the Executive Committee and the Conference Committee as well be fully advised and conferred with from time to time during the progress of the work, and

"FURTHER, It is the opinion of those present that the services of Joseph Dain, Jr., who has had wide experience in the development of the tractor in connection with the work of his father should be made use of to the fullest extent and in as important a capacity as his experience and abilities permit.

"December 11, 1917. Construction has been provided for 100 tractors and motors have been contracted for. Mr. Clausen stated that he expected 50 of these tractors would be completed by June 1, 1918."

Deere Buys Waterloo Company

On March 14, 1918, Deere & Company bought the Waterloo Gasoline Engine Company, makers of the Waterloo Boy tractor.

> "July 16, 1918. Mr. Morgan discussed at some length the tractor situation and stated that, in his opinion, we should not discard the Dain machine permanently. This tractor, in Mr. Morgan's opinion, has many good features and at a later date it might be advisable for us to continue the manufacture of this machine."

One hundred Dain tractors were built at the Tenth Street factory in East Moline, the same factory in which the first John Deere binders were built. Elmer McCormick was in charge of building the 100 tractors. He had been closely associated with Mr. Dain in working on the design of the tractor and engine almost from the start.

These tractors were built in 1919, and all were sent to the Huron, South Dakota, territory. There are no records to be found that give any information relative to the performance or disposition of these last tractors. Illustrations, drawings, and sales brochures of the Dain tractor were prepared as an aid to sell the tractors.

After this lot of tractors was built, there was no further development of the Dain tractor. It may be presumed that the following reasons contributed to the decision to stop further activities:

Why Dain Tractor Was Discontinued

1. In 1918, most of the full-line implement companies were building tractors, particularly International-Harvester Company and Case. It was felt that it was imperative to get into the tractor business at as early a moment as possible in order for Deere & Company to hold onto their plow business. A tractor and plow were usually sold by the same dealer.
2. That the Dain tractor, while a good tractor, was high-priced and also that it would require considerable time to tool up for manufacture.
3. That Mr. Joseph Dain, Sr., who was responsible for the Dain tractor, had died.
4. That Deere & Company had purchased the Waterloo Gasoline Engine Company and thus had a factory, a tractor, and an organization that was functioning, and so was in the tractor business at one quick stroke.

In concluding this chapter on the Dain tractor, it should be emphasized that in his thinking about tractors Mr. Dain was ahead of his time, for he insisted that tractors must be made better in every way, and that price was not the main objective. Many features of the Dain tractor were better than those of most tractors of that period. ■

DAIN TRACTOR SPECIFICATIONS

RATING: Drawbar, 12 hp; Belt, 24 hp.

SPEEDS: Two forward and two reverse. All speeds direct. No intermediate used. High, 2-5/8 mph; low, 2 mph. Reverse speeds same as forward.

WEIGHT: 4600 pounds.

LENGTH: Wheelbase: 6 feet 2-3/4 inches. Overall: 12 feet 6 inches.

HEIGHT: 4 feet 9 inches.

WIDTH: 6 feet 4 inches.

WHEELS: All traction. Rear, 40 inches in diameter, 20-inch face; front, 36 inches diameter, 8-inch face.

STEERING GEAR: Worm and sector.

TRANSMISSION: Exclusive design. No gears to shift in changing speeds. Change can be made from low to high or vice versa, while under full load without stopping the machine. All gears are cut from special heat-treated and tempered steel mounted on ball and roller bearings and enclosed in an oil-tight case. Chains are of highest grade steel and are of the roller type.

SPROCKETS: Cut from heat-treated and tempered steel.

MOTOR: Size: 4-1/2 inches bore by 6-inches stroke. John Deere design. All bearings extra large. Large hand-hole plates through which the bearings may be adjusted or the pistons removed.

CYLINDER: Cast-in-block and ground to size. Detachable head. Extra-large water jackets.

PISTONS: Extra long and ground to size.

PISTON PIN: Extra large, 1-1/2 inches in diameter, fastened in the piston. Bearing in connecting rod head.

CONNECTING RODS: Drop-forged from open-hearth steel and double heat-treated.

CRANKSHAFT: Three bearings, 2-1/2-inches in diameter. Drop-forged from open hearth steel and double heat-treated.

BEARINGS: Connecting rod, 2-1/2" diameter by 2-7/8" long. Front main, 2-1/2" diameter by 3-7/8" long. Center main, 2-1/2" diameter by 3" long. Rear main, 2-1/2" diameter by 4-1/8" long. All bearings bronze-backed with special Babbitt metal lining.

LUBRICATION: Force-feed from mechanical oiler, supplemented by a splash system.

GOVERNOR: John Deere design, built in.

COOLING: Water circulated by positively driven centrifugal pump of extra large capacity.

FAN: 19-inches in diameter. Positive gear drive.

CARBURETOR: 1-1/4-inches Model M-2 Stromberg.

AIR CLEANER: Bennett.

IGNITION: K-W high-tension magneto with starter coupling.

PULLEY: Steel, split hub, 30" diameter with 8" face. Belt speed, 2190 feet per minute.

Chapter 3 — The Sklovsky Tractor

In the spring, summer, and fall of 1915, there are numerous entries in Theo Brown's dairy in reference to a small tractor being designed under the supervision of Max Sklovsky, head of the Deere & Company Engineering Department. At this same time Joseph Dain was developing a tractor of his own design with power to handle a three-bottom plow. George Mixter was the one who directed Sklovsky to undertake the design of a two-plow tractor, as there was a feeling that there was a field for a two-plow tractor in addition to a three-plow tractor.

E.R. Wiggins, George Pearce, and Nathan Lesser, all in the Deere & Company Engineering Department, worked out the design of a small tractor under Max Sklovsky's supervision.

One-Piece Cast-Iron Body

A special feature in the design of this tractor was making the whole body, including engine pan, in a single casting. This is without doubt the first tractor to be designed as a solid unit.

Another feature was to so design this unit that there were no recessed surfaces involving costly machining. This contributed to a simple, strong construction.

Mr. Sklovsky's first design, called the Model A-2, was a three-wheel tractor, all wheels being power driven. There was no differential and the front axle was pivoted in the center like a wagon axle.

The A-2 Tractor was built at the Marseilles plant (now the John Deere Spreader Works). A Northway four-cylinder engine was used. This tractor was first tried out in the field on November 20, 1915. Items in Theo Brown's diary of 1915 tell of being in the field with this tractor and plow until the ground froze December 15. The outfit did pretty well, except that steering the tractor was almost impossible, since the tractor operator had to overcome the torque of the engine. This was due to having the power-driven front wheels mounted on a wagon-type axle and with no differential.

The tractor performance, other than steering, was satisfactory enough to warrant rebuilding and incorporating a pivot-axle, automobile-type of steer. This revised tractor, known as Model B-2, was built at the Marseilles plant and put to field work. On July 13, 1916, George Mixter reported to the Board of Directors as follows:

"Two-Plow Tractor

"One machine with a cast-iron frame completely enclosing all working parts has been built. This has a four-cylinder Northway engine and in general layout of driving all the wheels is a small edition of the Dain machine. Has been working continuous for about eight weeks, having plowed about 70 acres, pulled a drag for test purposes and recently has been cutting hay. Seems to contain one or two good elements for a small machine. Probably could be built with a suitable engine so as to be decidedly acceptable to the farmer.

"Cost figures, however, of $521 (2.25 steel) which figure is based on paying $140 for the engine, remains a commercial question. It is probable that on a manufacturing basis with proper development of the engine and a reasonable normal price for steel that the cost of manufacture would be just over $400, certainly not in excess of $450, which latter figure would mean $900 to the farmer.

"Nothing further has been done perfecting this particular type of small tractor because of the question in the writer's mind of the commercial practicability of a high-grade four-cylinder machine for two-plow work. Four-cylinder motors of this small size, running at speeds of 1000 rpm, apparently are not considered practical for burning kerosene, which is apparently a necessity.

"One-Cylinder Engine

"Some work has been done on the design of a two-plow tractor with a single cylinder engine, all-wheel drive, cast-iron frame. There seems to be sufficient possibility in this design to carry it somewhat further. Low cost, easy accessibility, certainty of burning kerosene being some of the desirable features.

"It is probably true that the inclination of the farmer is to buy a tractor that is as much like an automobile as possible. On the other hand, offering an all-wheel drive machine might overcome this point of view.

"Our efforts on this class tractor have thus far aimed at a durable and satisfactory machine at manufacturing cost of not over $300, i.e., $600 to the farmer."

War Stops Development

The Model B-2 tractor was built, and performed quite satisfactorily in the field. However, further development of this tractor was halted because of the more urgent war work demands on the small Engineering Department of Deere & Company and was never revived.

Max Sklovsky endeavored to build a tractor using farm implement construction as far as possible, so that the cost to the farmer would be low. ■

Page 31: *No photographs of the Sklovsky Model A-2 Tractor are known to exist, but this Engineering Department drawing, dated November 16, 1915, shows the A-2 with two-bottom plow attached. The Model A-2 had power-driven front wheels mounted on a wagon-type axle, with no differential, and was almost impossible to steer. It was followed by the much-improved Model B-2.*

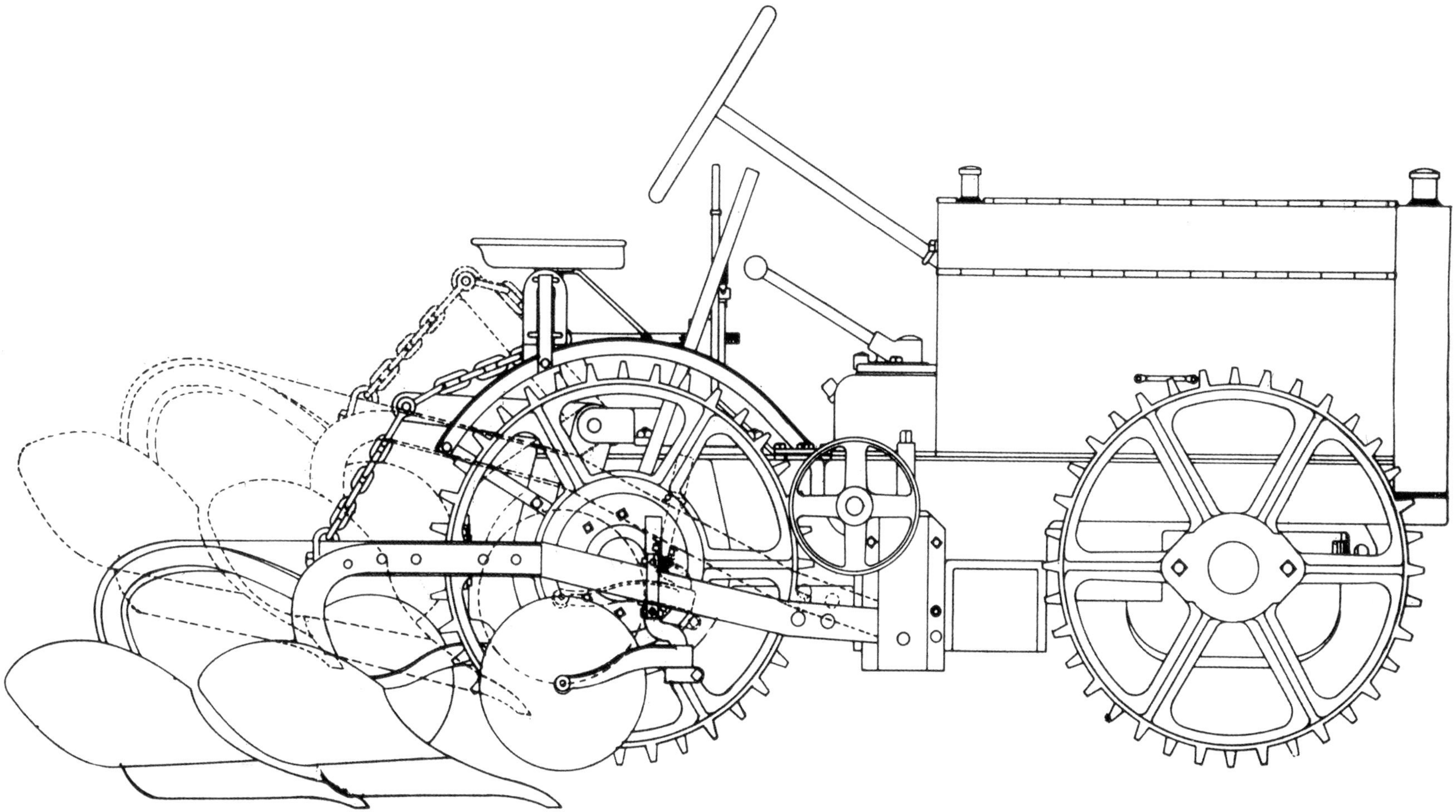

Tractor by Eng Dept
HARV STEEL

Page 32: Max Sklovsky on the Model B-2 Tractor, which had a four-cylinder Northway engine. The photo was taken next to the Harvester Works steel shed.

Right: Detailed photo of the Northway engine.

Page 34: Max is taking a ride through some standing water, which beats sitting in the office. This photo provides a right-side view of the Model B-2.

Right: The proposed Model D-2 Tractor would have brought several modifications over the B-2, but development was brought to a halt by the first World War and was never resumed.

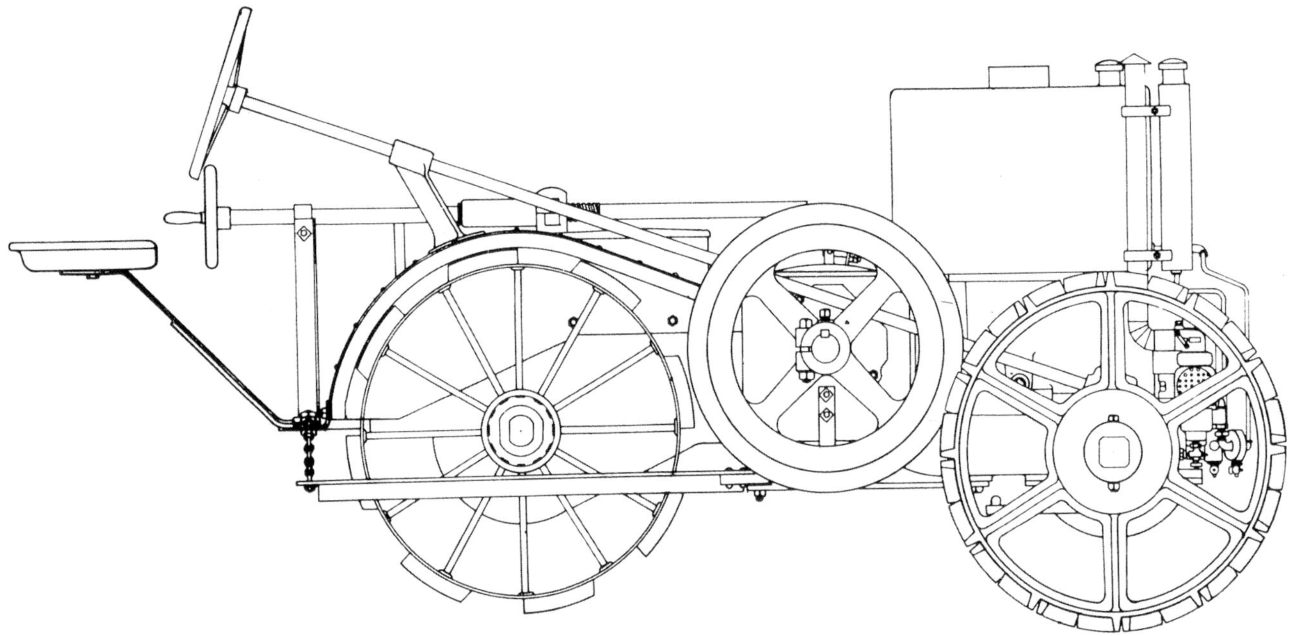

Chapter 4 — The Single-Row Motor Cultivator

The first attempts to design a tractor for cultivating row crops followed the pattern established for horsedrawn cultivators, except that a steering wheel replaced the harness reins. On both horsedrawn cultivators and early motor cultivators, the operator steered the outfit with his hands and manipulated the cultivating rigs with his feet.

A troublesome difference between horses and tractors immediately became obvious.

On horse-drawn cultivators, the operator's double job was easy, for horses soon learned to follow the rows, permitting the operator to give his full attention to dodging the rigs.

But if tractors didn't grab an occasional mouthful of corn leaves, neither did they learn to follow the corn rows. So cultivating, which had been a comparatively easy job with good horses, suddenly became tough, exacting work with the early motor cultivators, for the operator had to perform two unrelated functions simultaneously — looking ahead and steering the tractor, and looking down and manipulating the rigs.

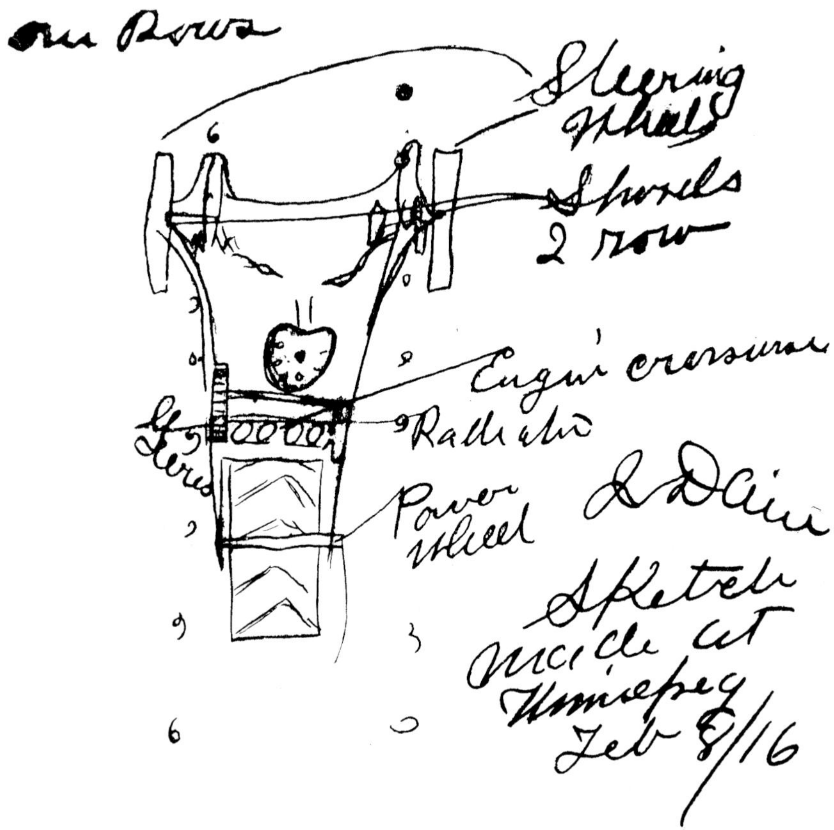

Examples of this type of tractor or motor cultivator were the Moline Universal and the Avery. In both of these outfits there was no coordination between steering and cultivating. This type of motor cultivator proved to be impractical, because a tractor operator could not successfully coordinate two unrelated functions.

Several Hit on Same Answer

United States Patent Office records show that early in 1916 there was considerable activity, by several individuals working independently of each other, on basic changes in motor cultivator design.

It is rather a coincidence that Ronning & Ronning, E.A. Johnston, Joseph Dain, and Theo Brown all conceived the idea of a motor cultivator in which a pivot-axle type of horse-drawn cultivator, with laterally swinging rigs, formed the front portion of the tractor. In this construction the steering of the tractor and the shifting of the cultivator rigs were accomplished with one and the same operation.

J.B. O'Donnell of Sheldon, Iowa, seems to have been the first to place a pivot-axle cultivator ahead of the tractor, but he did not use laterally swinging rigs. He filed his patent application July 14, 1915.

Mr. Dain's Historic Sketch

The first record of Deere & Company in the development of the motor cultivator is a sketch by Joseph Dain dated February 8, 1916. This sketch *(left column)* is preserved in the Deere & Company Patent Department files. Patent application was made June 6, 1916, and Patent No. 1,667,843 was issued May 1, 1928. This structure was never built.

The next activity was on March 30, 1916. An entry of that date in Theo Brown's diary reads: "Dr. W.E. Taylor suggested that we work on a cultivator pushed by a tractor. I sketched it out." This sketch is shown on page 37, and the Deere & Company Patent Department notes are reproduced below.

"Deere & Company

"March 30, 1916.

"Mr. Theophilus Brown submits a sketch of a cultivator and tractor coupled together, the object of the structure being that the control of the cultivator accomplishes the guiding of the tractor.

"Tractor cultivators up to this time have always had the tractor situated in front of the cultivator. The guiding being accomplished by the driver steering the tractor proper. This had disadvantages in that the cultivator does not respond as quickly to the guiding of the tractor as it should to do good cultivating, and the operator in watching the guiding of the tractor does not have time to observe the work and control the cultivator as he should.

"In the structure proposed by Mr. Brown the action of guiding the cultiva-

O522-10M-2-16

DEERE & COMPANY **MEMORANDUM**

Theo Brown

Witness — W G Duffield

MARCH 30 '1

tor gives immediate response to the cultivating parts as in the ordinary operation of a cultivator, and simultaneously the tractor is caused to follow in the same direction; in other words the single action of the operator in guiding and operating the cultivator as it should be operated also controls the travel of the tractor.

"In the sketch submitted it is proposed to use a K.C. cultivator which has pivot wheels and is so arranged that the gangs shift laterally in conjunction with the wheels. When it is desired to turn at the ends or when travelling on the road, the pivot wheels are to be locked or stiffened and the pair of wheels as an entirety swung about a central pivot on the frame. It is contemplated as a possibility to so rearrange the cultivating rigs that the pivot wheels can be used for steering at the ends on the road.

"The real feature of the basic thought as exemplified in the sketch, is the control in respect to guiding of both cultivator and tractor by a single action of the operator."

This reproduction of Deere & Company Patent Department notes explains two important principles of the Brown tractor cultivator — placing the power unit and drive wheels at the rear, and linking the front pivotal wheels to the moveable rigs to combine the two operations of steering and dodging while cultivating.

Brown's First Cultivator

Work was started immediately to build a motor cultivator following the ideas shown in the sketch. It was a matter of adapting a K.C. cultivator to be pushed by a power unit, consisting of a 7-1/2 hp New Way air-cooled engine, mounted on a truck supported by manure spreader wheels and using spreader drive sprockets and chain. On June 1 this outfit was used to cultivate corn with a considerable measure of success, and it was contin-

Left: *Overhead view of the Brown one-row cultivator. The operator steered the unit in the field by dodging the rigs, which also pivoted the front wheels. The steering wheel, which controlled the pivot joint between front and rear trucks, was used for turning at row ends.*

Above: *Gus Bischoff operating a Brown cultivator in July 1916. This outfit was used throughout the 1916 cultivating season with a considerable measure of success.*

Page 39: *This photograph was taken in early July 1916, when the Deere & Company Washington, D. C. patent attorney, H.H. Bliss, came to Moline to get first-hand information about patentable features of the Brown motor cultivator. Left to right are: Dr. W.E. Taylor, Theo Brown, Mr. Bliss, Gus Bischoff, George W. Mixter, Floyd R. Todd, Joseph Dain Sr., and Harold B. Dinneen.*

ued in use throughout the cultivating season. Minor changes were made from time-to-time to improve performance.

Early in July, H.H. Bliss, our Washington patent attorney, came to Moline to get a better understanding of the features that might be patentable in this motor cultivator.

On July 13, 1916, George W. Mixter reported to the Board of Directors on the subject of the tractor cultivator as follows:

"A machine built by Mr. Brown which places the drive wheel at the rear with the cultivator in front has been working for about a month. Its principal features are that one dodges corn by pivoting the front wheel with one's feet and at the same time dodging the rigs. The cultivator, in fact, being an ordinary two-row cultivator reduced to a single row. Combination is a pivoted joint between the front and rear members controlled by a steering wheel, which is used principally for turning at the end of the row.

"Engine Was Big Share of Cost

"The machine in the field was hastily constructed from available parts and is driven by a 7-1/2 hp New Way engine. It has proven distinctly a success and very attractive. The machine, other than the engine, weighs about 700 pounds and will cost about $28 (1.25 steel), or $35 (2.25 steel). The New Way engine was billed to us at $115, but this price is probably excessive. It looks as though the machine could be sold to the farmer for $300, and ultimately for $250. (Note: The thinking then was in terms of farm implement construction and design.)

"It is probable that applications of Mr. Dain and Mr. Brown for patents will ultimately involve quite complete patent protection on the particular combination this machine rep-

Above Left: *One of the 25 one-row "Tractivators," equipped with McVicker two-cylinder hopper-cooled engines, built at the Marseilles plant (now John Deere Spreader Works) in East Moline in 1917. Tested throughout the middle west, they did a good job and were easy to operate — but unfortunately did only about as much work in a day as a man with a team and horsedrawn cultivator.*

Above and Page 41: *While McVicker was designing a new engine for the "Tractivator," the John Deere Plow Works was redesigning the tractor frame to simplify manufacture, and also with the idea of removing cultivating rigs and running the tractor "backward" for other work, such as plowing, disking and mowing* (see photos on page 43).

resents; possibly the protection can be made quite broad. The Plow Department is at once going ahead with a carefully designed machine with a single cylinder New Way motor, meanwhile other motors are being investigated."

Ronnings Win Patent Race

In reference to patent protection mentioned by Mr. Mixter, it developed that in 1925 Ronning & Ronning were granted the basic patent on the conception of a motor cultivator in which steering the tractor and laterally swinging the cultivator rigs were one and the same function. Their patent application was dated March 6, 1916; Joseph Dain's, June 6, 1916; Theo Brown's, July 10, 1916; and Ed Johnston's, October 25, 1916.

It is of interest to note that Ronning's patent was issued in 1925, the same year that International-Harvester sold their first 250 Farmall tractors. The Farmall tractor equipped with cultivator was held to infringe the Ronning patent, so International-Harvester paid handsomely for its use — $1 for every tractor, whether or not equipped with cultivator.

Six patents were issued to Deere on its 1916–1917 motor cultivator efforts, Nos. 1,441,476; 1,426,544; 1,441,480; 1,451,672; 1,476,687; and 1,667,843.

More Powerful Engine

It was found necessary to provide an engine for the motor cultivator with more power than that of the New Way air-cooled engine used on the first experimental outfit. The Associated Manufacturers Co. of Iowa was contacted and their engineer, Theodore Menges, designed a new engine for the job which when built did not prove a success.

The next step was to retain McVicker, a consulting engineer, of Minneapolis to design a twin-cylinder, hopper-cooled engine. This engine was built for Deere & Company by the Associated Manufacturers Co. At the same time the Plow Works was working on a complete redesign of the whole tractor outfit to make it suitable for manufacture.

During this period of development the possibility of using other implements with the tractor was advanced to the point of actual field work. Removing the cultivator rigs and running the tractor backwards made a general-purpose tractor of the job. How this was accomplished can be seen in the photos on page 43.

These various implements worked with some measure of success. On September 12, 1916, W.R. Morgan, Manager of the Harvester Works, reported to the Board of Directors that "the original tractor cultivator has been taken to the Harvester plant, and that he had attached the machine to a mower and tried it out quite successfully."

Dinneen Reports Progress

A complete redesign of tractor and engine was completed in December 1916. On January 16, 1917, H.B. Dinneen, manager of the Plow Department, reported to the Board of Directors on the progress of the tractor cultivator to date as follows:

"1. The machine other than the engine had been carefully worked over and is apparently ready to manufacture subject to Texas trials.

"2. A two-cylinder, hopper-cooled engine designed by McVicker has been built by the Associated Manufacturers of Waterloo.

"3. Tests of the engine as well as suitability and design were satisfactory, except:

"4. That the water consumption at three gallons per hour at steady full load when running continuously in the shop, seemed excessive.

"5. That the time available did not permit further major changes in engine design for machine to be furnished the spring trade. After reviewing the matter at length it seemed probable to the Board that the engine would be satisfactory for cultivating, and that the water consumption would be rather less under field conditions than during the shop trials."

Board Orders 25 Tractor Cultivators

The Board passed the following resolution:

> "That the recommendation of Mr. Dinneen that material for 50 machines with the two-cylinder engine be provided and that 25 machines be built as rapidly as possible, it being understood that two machines are to be tried in Texas at the earliest possible date, and the continuance of work on the 25 machines and the possible completion of 50 machines, be again considered after the Texas trials."

Work on the 25 machines was carried on at the Marseilles plant. (Note: In those days there was not the engineering knowledge nor the accuracy in manufacture that exists today, so as a result the troubles and difficulties in building the 25 outfits were many and the finished product would in no respect meet today's standards.)

On February 17, 1917, the first tractor was completed, and photographs were taken on February 22.

Stone-Boat Endurance Test

On March 10 two motor cultivator outfits were started on an endurance test, dragging stone-boats in the muddy field behind the Marseilles plant. Conditions made it possible to conduct the tests there instead of going to San Antonio.

On March 14, George Mixter reported to the Deere & Company Board of Directors:

"We expect to have 25 machines available by the 15th of April. There have been two machines in the field here and they have shown some pretty good characteristics. The engine pulls to a degree he did not think was possible, and the proposition is looking better right along. He said that on the basis of costs recently made up the tractor would probably have to be sold to the farmer for $475. He said that the work of getting out 25 machines is being pushed very vigorously, and if we come through with the proposition as well as we hope, we are going to have a problem immediately in front of us for the production of these machines for sale a year from this spring."

On March 7, 1917, letters had been sent to the following Branch Houses: Moline, Omaha, Minneapolis, Kansas City, St. Louis, and Indianapolis which said in substance,

> "The motor cultivator has now reached a stage of development where we want to put out 25 machines to work continuously cultivating corn during the spring and summer; and after the corn is cultivated, cutting hay." Each Branch House listed was assigned a definite number of tractors.

New Machine Named "Tractivator"

On April 10, L.R. Clausen was put in charge of the motor cultivator (now called Tractivator) program.

The 25 Tractivators were built a the Marseilles plant and sent to the Branch Houses as designated. All of these outfits were used for cultivating corn, and some were later used for cutting hay. Special factory men followed these Tractivators and made daily reports. A complete set of the reports is in the Patent Department files, and a careful study of these reports gives a rather complete picture of the field performance of these 25 Tractivators and the reactions of the men who used them.

The reports clearly indicate that the Tractivator as a cultivating outfit performed satisfactorily both as to quality of work and ease of operation. Thus, the basic idea of shifting the cultivator rigs in dodging corn and steering the tractor with one and the same operation was proved to be sound.

However, the reports revealed that a man with the Tractivator did not accomplish any more work in a day than he could with a pair of horses, and for that reason he could not afford to buy the Tractivator. A study of the reports have the answer as to why the outfit failed. The reasons listed were as follows:

1 Lack of sufficient power to work on hilly ground.
2. With a one-speed transmission the tractor ran too slow for most conditions, and this was aggravated by the lack of engine power.
3. Excessive evaporation of water in the hopper-cooled engine, amounting to as much as two gallons an hour. It required time and effort to replace the water.
4. Leaky carburetors gave a low efficiency engine.
5. Excessive consumption of lubricating oil, due partially to poor workmanship, was both expensive and annoying.
6. Added to the above complaints was the fact that many of the Tractivator users had heard of, and some had seen, the new International-Harvester motor cultivator which was a two-row cultivating outfit enabling a man to do twice the work of a one-row. This International Motor Cultivator was brought out in 1917, but was subject to upsetting due to the power plant being mounted above the rear drive wheel. It was not commercially successful. The small-diameter drive wheels left deep and objectionable tracks which the front-mounted rigs could not smooth out.

Single-Row Idea Dropped

In view of all the reports on the Tractivator, the 25 outfits were returned to the factory and all activity on the single-row idea was dropped. In 1916, it was thought a motor cultivator should be of implement construction, rather than of automotive design and precision manufacture. Had the tractivator been as well-engineered and as well-built as present day one-row cultivating tractors the one-row idea in 1916 would have found a more ready acceptance. ■

Chapter 5 — The Two-Row Motor Cultivator

By the middle of June 1917, reports from the field on the 25 one-row Tractivators had pretty well established three important facts, which were:

1. One-row motor cultivators probably would not be economically successful, for farmers were finding that they could cultivate just as much ground in a day with a team of horses.
2. Row crops could, however, be successfully cultivated with mechanical power.
3. Therefore the logical step would be to develop a two-row motor cultivator.

Silver's First Cultivator

Using an Avery motor with friction transmission, and such other available parts as he could find, Walter Silver built a two-row motor cultivator in two weeks' time. A photograph taken June 30, 1917, shows Mr. Silver cultivating corn at the Marseilles plant with his first experimental machine.

Silver's first cultivator had power-driven — but not steerable — cultivator wheels. A small rear wheel was steerable. The operator shifted the cultivator rigs laterally with his feet, and as a separate operation steered the outfit by means of a hand wheel which controlled the rear tractor wheel.

As with previous motor cultivators, it soon was evident that requiring the operator to perform two unrelated fucntions — steer the tractor and manipulate the rigs — was not satisafactory. How Silver solved this problem is recorded in Mr. Brown's diary under date of August 9, 1917: "Went with Dain and Clausen to see a new motor cultivator at the State Farm. It is a two row machine built with power on the two wheels and these wheels are dirigible. There is a small wheel that can be steered to help in turning around. The drive chain was a log chain as it has to stand a twist."

This structure incorporated the principle of combining two functions — steering the dirigible front cultivator wheels, and laterally shifting the cultivator rigs — into one and the same operation by the tractor driver. The rear wheel was used only for sharp turns at row ends.

Method is Patented

The earliest record of such a combination is shown in a sketch dated January 30, 1917. An entry in Brown's dairy of that date reads: "Gave Duffield (in charge of Deere & Company Patent Department) a series of sketches showing all the different combinations I could think of for motor cultivators." A patent on this combination was applied for on March 23, 1917, and issued as Patent No. 1,451,672.

On September 12, 1917, there is this entry in the minutes of the Board of Directors' meeting:

> "Mr. Dain predicted that in five years there will not be half the horsedrawn implements that are sold today, unless they are adapted to tractors. We have got to have a motor cultivator to protect our cultivator trade. There will be 5000 motor cultivators sold next year. There is not any chance for use to get any part of that business unless we have the right kind of motor cultivator."

Silver was directed to continue development of the two-row motor cultivator. His base of operations was moved to the Tenth Street factory in East Moline, where the first John Deere binders were built. Here during the fall and winter of 1917–1918 a complete redesign of the motor cultivator took place. The first one built was photographed on April 15, 1918, and sent to J.B. O'Donnell at Sheldon, Iowa, for a thorough tryout on his farm.

Left: *Using an Avery four-cylinder engine with friction transmission, and other available parts he could find, Walter Silver built a two-row motor cultivator in two weeks' time.*

Page 45: *Walt Silver operating the two-row motor cultivator he built in June 1917, after it became apparent that farmers were not overly interested in one-row motor cultivators.*

Pages 46 and 47: *Front and rear view of Silver's first two-row motor cultivator. The operated shifted the cultivator rigs with his feet to dodge the corn, and steered the cultivator by means of a hand wheel which controlled the dirigible rear wheel — two distinctly separate operations.*

O'Donnell Also An Inventor

O'Donnell was a farmer with a mechanical turn of mind. It is probable that he was the first one to push a two-row cultivator with mechanical power. Several photographs indicate that O'Donnell connected a Bull tractor to a two-row horsedrawn pivotal wheel cultivator so that the tractor pushed the cultivator, and the operator steered the outfit by foot pedals. One difficulty that made the outfit impractical was that there was no provision for short turns at the end of the row.

Deere & Company had come in contact with O'Donnell on patent matters, and had employed him from February 1917 to November 1917, to follow the 25 Tractivators in the field and to make daily reports on their performance.

Silver's No. 4 Motor Cultivator

After this first redesign had been sent to O'Donnell at Sheldon, Silver made a few further changes and five more outfits were built. Photographs of this No. 4 model were taken on June 7, 1918, and then the outfits were sent to farmers for a workout under field conditions.

On June 11, 1918, L.R. Clausen reported that "five two-row motor cultivators had been made and were now working in the field with very satisfactory results. It would appear that there was some real value in the machine."

Again on June 18, Mr. Clausen stated "that the five motor cultivators which had been

designed by Mr. Silver were now operation in the fields, and were doing very satisfactory work. He considered that Mr. Silver had done a very fine job in designing these machines."

Again on August 13, Mr. Clausen said "that Mr. W.H. Silver, who has been handling the Motor Cultivators Works, has been assigned to the Plow Works as foreman of the Experimental Department. Mr. Silver will make a detailed report... at which time the motor cultivator situation will be brought up for discussion."

On August 15, 1918, Mr. O'Donnell reported on the performance of the motor cultivator as follows:

Left: *Silver's cultivator was completely redesigned during the winter of 1917–18, and the Model 3, shown here, was sent to J.C. O'Donnell of Sheldon, Iowa, for trial.*

Above *Front view of Silver's Model 3, which obviously offered the operator an unusually good view of the work being performed.*

Page 49: *The hand-controlled rear wheel was used for short turns at row ends; the powered front wheels were linked to the rigs, so that shifting the rigs also steered the outfit.*

JOHN DEERE PLOW
JOHN DEERE CULTIVATOR
JOHN DEERE PLOW

"REPORT ON MOTOR CULTIVATOR

"by J.B. O'Donnell

"Disked 175 acres in 110 hours, using 96 gallons of gas, 5-1/2 gallons of oil (motor and transmission) and 6 gallons of water. This work was done on spring breaking bluegrass sod, pulling an eight-foot disc harrow with about 275 pounds weight on the disc, and running the tractor in low speed.

"Planted 130 acres of corn in 43 hours using 46 gallons of gas, 6-1/2 quarts of oil and 4 gallons of water. Thirty acres of this corn was checked in with good results. We got fairly good cross check and had no trouble cultivating both ways.

"Cultivated 200 acres of corn in 79 hours using 74 gallons of gas, 4 gallons of oil, and 5 gallons of water. Twenty acres of this corn was cultivated crossways the second time through, driving the tractor in low speed. It was run 14 hours using 16 gallons of gas, 2 quarts of oil, and one gallon of water. The remaining 180 acres was cultivated both first and second times through (this corn was as high when cultivated the first time as corn ordinarily is when cultivated the second time, and was laid by with the second cultivation), running the tractor in high speed. Run 65 hours and used 6-1/2 gallons of gas, 3-1/2 gallons of oil, and 5 gallons of water.

"Three Miles An Hour On Binder

"Cutting grain pulling an eight-foot McCormick binder. Cut 30 acres in 15 hours using

15 gallons of gas, 2 quarts of oil and 2 gallons of water. The machine worked out very good on the binder, and by speeding up the motor we got a good working speed of about three miles an hour with the tractor in low gear. We were easily able to make a square turn with the binder by releasing the rear wheel lock and applying the brake. It was not necessary to stop or slow down for the corners. The grain we were cutting was lodged badly and it was necessary to make frequent stops with the binder.

"We also hitched the machine to a twenty-foot drag harrow and it worked out good, but we did not do very much work with it as our corn was on spring breaking and we did not drag it after the planter.

"We have done a good deal of work with this machine, and the machine worked out very successfully on each of the jobs mentioned above. We have had no trouble at all with the motor or transmission gears. The motor has never refused to run, or given any trou-

Left: *The Model 3 equipped for planting*

Above: *Note how depressing the right foot pedal steered the powered front wheels to the right.*

Page 51: *Mr. O'Donnell used the Model 3 on an 8-foot McCormick binder, as well as on a disc harrow, 20-foot drag harrow, road drag, corn planter, corn cultivator, and mower.*

ble in starting. In fact, it has required no special attention whatever; not even a spark plug has been removed. The whole machine is practically in as good condition now as when it left the factory.

"More Power Needed

"Our main drawback all the way through has been the lack of power.

"You will note that in disking we only averaged about 16 acres per 10-hour day, where if we had been able to pull a ten-foot disc harrow on high speed, we could do almost double that amount. In planting corn we pulled the planters on high speed practically all of the time, but occasionally would have to shift to low speed.

"In cultivating corn we did most of the cultivating on high speed, but did not cultivate quite so deep as we would like to have done had we the power to pull it.

I called on J.A. Kroon, Pacific Junction, Iowa, some time ago. He was pulling two five-foot mowers and they worked out very good on our machine. He was very much pleased with the machine and said he was in the market for one as soon as it could be delivered to him.

"I also called on Mr. R.L. McPhersen at Tarkio, Missouri. He had cultivated 69 acres of corn the second time through, running 43 hours and using 42 gallons of gas, 2 gallons of oil, and 25 gallons of water. Mr. McPhersen said the machine worked out fine in every way, and he had no trouble with it on level ground, but did not have enough power to work the hills.

"Design Considered Successful

"Our experience this season has proven our present design to be very successful, not only in cultivating corn, but in handling all farm implements so far as our experiments have gone.

"We have already used the machine on the disc harrow, drag harrow, road drag, corn planter, corn cultivator, mower and binder, and there is no reason why it will not work out just as well on the two-bottom plow, corn husker, and other implements as it did on the above-mentioned tools, besides the many different jobs it will do on the belt.

"When I say it worked out on those implements, I mean that all of those implements worked out just as well on our machine as they do on any other tractor, and we can go into the field in competition with any other machine and prove the truth of these statements.

"Much More Practical...

"After having spend the season in the field with this machine and also having had the

Above Left: *O'Donnell reported that the No. 3 "worked out good" when hitched to a 20-foot drag harrow.*

Above: *Barely had a No. 3 been sent to O'Donnell for testing before Silver incorporated several improvements and modifications (particularly in the design of the power-driven cultivator wheels) in his No. 4, of which five were built and sent to various farmers for testing. The one shown here was the only one of the five that was known to be decorated with pinstriping.*

Page 53: *Rear view of a No. 4.*

JOHN
DEERE
CULTIVATOR
DEERE
PLOW

opportunity to be in a field with several other tractors, such as the Fordson, Avery Motor Cultivator, International Motor Cultivator, Moline Universal and others, I feel confident that our design is right and that our machine is much more practical than any other machine on the market at the present time.

"We can plant corn and cultivate corn far better and do many other jobs just as well as any other tractor, and we do not have any complicated hitches for attaching the tractor to the implement we are using, nor do we have to have a special-built implement strong enough to carry the weight caused by the torque of the motor as this weight is always carried on the rear castor wheel. We must have at least 6-hp motor on this machine, and when the job is finished, the proper hitches worked out for the different implements, it will be in reality a universal machine.

"Farmers Wanted Light Tractor

"There are more farmers in the market today for this type of tractor than any other machine of its kind. I have talked to hundreds of farmers in the past two years who are interested in a light tractor, and I would seldom fail to ask a farmer what tractor he believes would be best adapted to his needs and the most practical for him to buy, and I believe I am safe in saying that 50% of them would mention a certain tractor that is advertised as a universal machine with a large number of different farm implements. I only mention this to show that the farmer is interested in and looking for a more universal type of tractor than he has been able to get so far, and it is not hard for anyone to understand why a machine that can be used practically every month in the year is more practical than one that is only used for a few jobs and must stand idle the rest of the time.

"In designing this machine, it should be remembered that it is not only a motor cultivator, but an all-purpose tractor, and that the farmer will have work for it to do practically every month in the year.

"For these reasons, it should be built of the very best material and equipment that is practical for a machine of this kind. I believe we should have Hyatt roller bearings throughout, or some other anti-friction bearing equally as good.

"$1000 A Fair Price

"The farmer has had enough experience with a cheap tractor and if we can show him that the John Deere tractor will give good service as John Deere implements have always done,

Above Left, Above, and Page 54: *The No. 4 in the field. The hood has been removed, probably to allow engine heat to escape.*

he will not hesitate to pay a fair price for the machine. I would consider $1000 a fair price for this machine under present conditions.

"Respectfully submitted,

"J.B. O'Donnell"

On September 3, 1918, Mr. Clausen presented a detailed report of the operation of the five motor cultivators to the Board of Directors:

> "These tractors have been in the field the past season and I recommend that the five machines in their experimental stage be used another year, with such modifications and changes as experience indicates advisable, under the continued charge of Mr. Silver."

On December 23, 1919, it was decided to transfer the motor cultivator development work to the Plow Works.

Experimental One-Row Machine

During the spring of 1921, an experimental single-row motor cultivator was built. In its construction it followed the ideas incorporated in the first two-row cultivator of 1917. The cultivator rigs were shifted laterally by foot pedals, and the tractor was steered by guiding the two rear caster wheels. Thus, there were two distinctly separate functions for the operator to perform. This double function idea was a step backwards. There are only three references to this project and they are found in Theo Brown's diaries, and are as follows:

> "June 8, 1921. The single-row motor cultivator went to the Einfeldt farm today. Weighs 1359 lbs.
>
> "June 9, 1921. The motor cultivator started working today. Mr. Butterworth saw it.
>
> "June 13, 1921. George Mixter was here and went out to see the motor cultivator and new Waterloo tractor."

On October 25, 1921, Mr. Clausen presented some photographs of the motor cultivator and indicated there is some question whether or not any further money should be spent in the development of the same in 1922. It was the sentiment of the Board that we discontinue any further development of the motor cultivator for one year.

Further Development Stopped

As result of action at the Deere & Company Board of Directors' meeting in October 1921, all further development on motor cultivators was stopped. There were several reasons for such an action.

1. The years 1921–22 were disastrous ones for the farm equipment industry. Every expenditure was scrutinized closely and only the most urgent was given any consideration.
2. While considerable experimental work was carried on, the Company's whole tractor effort was focused on the redesign of the Waterloo Boy Tractor.
3. The two-row motor cultivator design was not adaptable for plowing. The wheel tread was too wide and not adjustable.

The time had arrived when the motor cultivator was expected to be a general-purpose tractor, and, to be acceptable, had to be a plowing tractor too. The necessary flexibility was lacking in the design of the motor cultivator to make it satisfactory for plowing, although the tractor had performed other farm operations quite successfully.

The General-Purpose Tractor

It may be of interest to trace the development of the general-purpose-type of tractor today. The general-purpose tricycle-type tractor is an outgrowth of the two-row motor cultivator of 1920. At that time there were two schools of thought as to how a motor cultivator should be designed. One group believed the cultivator should be in front, the other was equally sure it should be behind.

The front-mounted cultivator with dirigible power-driven cultivator wheels was easiest for the operator. The rear-mounted cultivator gave the operator two functions to coordinate — one to steer the tractor, the other to shift the cultivator rigs. This was not an easy task.

The next step, which came not long after, was to take a motor cultivator in which the cultivator was mounted at the rear (as in the Rock Island) and make two alterations. 1st: Take the cultivator rigs from the rear and attach them to the tractor some distance ahead of the rear wheels. 2nd: Make an adjustable rear axle tread. That, in essence, made the general-purpose tractor as we know it today. ■

Above: and Page 57: *J.A. Kroon of Pacific Junction, Iowa, was "very much pleased" with the No. 4 pulling two 5-foot mowers. (The glass negative for page 57 was broken.)*

Left, and Page 59: During the spring of 1921, an experimental one-row motor cultivator was built. Its construction followed the ideas incorporated into the first two-row cultivator of 1917. These photos of the completed outfit were taken June 7, 1921, the day before it was taken to the field for testing.

Left, and Pages 61, 62, and 63: These photos, taken two weeks after the "studio" photos on pages 58 and 59, show the one-row motor cultivator at work in the field, and mark the end of an era insofar as Deere & Company's developmental work on this type of machine. Focus was then shifted towards the redesign of the Waterloo Boy Tractor, and the development of a general-purpose tractor. From those efforts came the Model "D" and Model "C" Tractors.

JOHN DEERE

APPENDIX

Bischoff, Gus	Farmer at Deere & Company Experimental Farm (where Molette addition to Moline now is located).
Bliss, H.H.	Deere & Company patent consultant in Washington, D.C.
Butterworth, William	President of Deere & Company, 1907–1928. Chairman of Board, 1928–1936.
Clausen, L.R.	Deere & Company Vice President in charge of manufacturing, 1919–1924 (succeeding George W. Mixter); in 1924 elected President of J.I. Case Company.
Dain, Joseph Sr.	Founder of Dain Manufacturing Company, which became a part of Deere & Company by merger in 1911; thereafter Vice President and member of Board of Directors of Deere & Company; Died in 1917.
Dain, Joseph	Now Assistant Secretary of Deere & Company.
Dinneen, Harold B.	Manager of John Deere Plow Works until 1919; Vice President of Minneapolis-Moline Company from 1919 until his death in 1937.
Hornburgh, C.H.	Division Sales Manager and later Sales Manager (upon retirement of George P. Shutz) of Deere & Webber Company.
Johnston, E.A.	Manager of Experimental Department, International Harvester Company.
Kirby, Fred	Superintendent of Marseilles Company (now John Deere Spreader Works). Now retired.
Lesser, Nathan	Engineer in Deere & Company Engineering Department; later succeeded Max Sklovsky as Chief Engineer of Deere & Company.
Livesay, R.C.	Experimental Department, John Deere Harvester Works.
Lourie, R.B.	Manager, John Deere Plow Company of Moline. Member of Board of Directors of Deere & Company. Now retired.
McCormick, Elmer	Worked with Mr. Dain on Dain tractor; later Chief Engineer of John Deere Waterloo Tractor Works until his death.
Melvin, C.H.	In John Deere Plow Works Experimental Department until transferred to Deere & Company to design tractor. Died in 1917.
Mixter, George W.	Member of Deere & Company Board of Directors, 1905–1942; Vice President in charge of manufacturing, 1907–1919. Great-grandson of John Deere.
Molstad, John	Territory Salesman, later Sales Manager, of Deere & Webber Company. Died in 1953.
Morgan, W.R.	Manager of John Deere Harvester Works; member of Board of Directors of Deere & Company from 1912 until his death in 1923.
McVicker	Consulting engineer located in Minneapolis, Minneosta.
Neighbour, L.B.	Then and now Manager of Experimental Department of John Deere Spreader Works (then Marseilles Company).
O'Donnell, J.B.	Farmer at Sheldon, Iowa.
Pearce, George	Engineer in Mr. Sklovsky's department.
Ronning & Ronning	Inventors living in Minneapolis, not connected with any implement company. Adolph Ronning invented the Roll-O-Matic front wheel construction used on modern John Deere tractors.
Schutz, George P.	Sales Manager, Deere & Webber Company, until retirement.
Silver, Walter	Engineer in (now Manager of) John Deere Plow Works Experimental Department.
Sklovsky, Max	Chief Engineer of Deere & Company. Now retired.
Taylor, Dr. W.E.	Head of Deere & Company Soil Culture Department 1910–1935.
Todd, F.R.	Formerly with Kemp & Burpee Manufacturing Company, which became part of Deere & Company by merger in 1911; Vice President of Deere & Company, member of Board of Directors of Deere & Company 1911–1933.
Velie, C.D.	Vice President of Deere & Webber Company; member of Board of Directors of Deere & Company 1911–1929. Grandson of John Deere.
Velie, S.H.	Manager of John Deere Plow Company of Kansas City; member of Board of Directors of Deere & Company 1911–1919. Grandson of John Deere.
Webber, C.C.	Vice President and member of Board of Directors of Deere & Company 1911–1944. President and Manager of Deere & Webber Company. Grandson of John Deere.
Wiggin, E.R.	Engineer in Mr. Sklovsky's department.

Appendix as provided in Theo Brown's book *Deere & Company's Early Tractor Development* was current as of 1953.